Regenerative Engineering Society

Held at the 2021 AIChE Annual Meeting

Boston, Massachusetts, USA and Online
7 - 11 November and 15 - 19 November 2021

ISBN: 978-1-7138-5688-7

Printed from e-media with permission by:

Curran Associates, Inc.
57 Morehouse Lane
Red Hook, NY 12571

Some format issues inherent in the e-media version may also appear in this print version.

Printed with permission by Curran Associates, Inc. (2022)

For permission requests, please contact AIChE
at the address below.

AIChE
120 Wall Street, FL 23
New York, NY 10005-4020

Phone: (800) 242-4363
Fax: (203) 775-5177

www.aiche.org

Additional copies of this publication are available from:

Curran Associates, Inc.
57 Morehouse Lane
Red Hook, NY 12571 USA
Phone: 845-758-0400
Fax: 845-758-2633
Email: curran@proceedings.com
Web: www.proceedings.com

TABLE OF CONTENTS

Recent Advances in Highly Biocompatible Zwitterionic Materials 1
Shaoyi Jiang

Designing Biomaterials for Three-Dimensional Tissue Bioprinting 2
Y. Shrike Zhang

Designing Engineered Tissue Platforms for in Vitro Disease Modeling and Regenerative Applications 3
Catherine Whittington

Biofilm Extracellular Matrix Protein: From Functional Amyloids to Versatile Biomaterials 4
Yichun Wang

Immuno-Engineered Biomaterials Reduce Implant-Related Infections 5
Bingyun Li

4D Printing: The Promise of Externally Controlling Internal Biomaterial Shape 6
Thomas Webster

Engineering in Precision Medicine 7
Ali Khademhosseini

Biomaterials that Enable and Drive Scaffold-Free Tissue Formation 8
Eben Alsberg

Engineering Targeted Therapeutics for Triple Negative Breast Cancer 9
Debra Auguste

Biodegradable Scaffolds for Large Peripheral Nerve Defect Regeneration: Electrical and Chemical Stimulation 10
Sangamesh Kumbar

Advances in Translation of SAPH Peptide Biomaterial-Drugs 13
Vivek Kumar

Recent Advances with Silk Proteins for Medicine 14
David L. Kaplan

Author Index

635239 Recent Advances in Highly Biocompatible Zwitterionic Materials

Abstract:

An important challenge in many biomedical applications is the prevention of unwanted nonspecific biomolecular and microorganism attachment on surfaces. Zwitterionic materials have emerged as a new class of biomaterials and have been demonstrated to highly resist nonspecific protein adsorption and microorganism attachment from complex media. In this talk, recent advances in the molecular understanding, design and development of zwitterionic materials will be discussed. Unlike industrial gold-standard poly(ethylene glycol) (PEG), there exist diversified zwitterionic molecular structures to accommodate various properties and applications. Newly developed zwitterionic materials will be illustrated. In addition, the new applications of zwitterionic materials to implants, stem cell cultures, medical devices, and drug delivery carriers will also be discussed. With zwitterionic coatings, hydrogels or nanoparticles, new capabilities have been achieved, including no capsule formation upon subcutaneous implantation in mice up to one year, expansion of hematopoietic stem and progenitor cells (HSPCs) without differentiation, no anti-coagulants needed for blood-contacting devices, and no antibodies generated against zwitterionic polymers conjugated to high immunogenic biologics.

621887 Designing Biomaterials for Three-Dimensional Tissue Bioprinting

Abstract:

Over the last decade, three-dimensional bioprinting has offered great versatility to fabricate biomimetic volumetric tissues that are both structurally and functionally relevant. It enables precise control over the composition, spatial distribution, and architecture of the bioprinted constructs, facilitating recapitulation of the delicate shapes and structures of targeted organs and tissues. In this talk, I will discuss our recent efforts on developing various cytocompatible, cell-instructive, and functional bioink formulations in promoting the fabrication of engineered tissue constructs, using a series of established or customized bioprinting strategies.

631892 Designing Engineered Tissue Platforms for in Vitro Disease Modeling and Regenerative Applications

Abstract:

Our research group uses biomaterial-based platforms and tissue engineering strategies to design *in vitro* models that capture the complexities of the changing tissue microenvironment in development, regeneration, and disease. One particular complexity of interest is fibrosis, which occurs during wound healing and disease. The fibrotic tissue microenvironment experiences increased extracellular matrix (ECM) deposition, crosslinking, and linearization of ECM fibers, while undergoing degradation over time. This dynamic ECM remodeling produces an altered ECM with heterogeneity in stiffness, architecture, and biodegradability at the micro- and macroscales. Yet, current preclinical *in vitro* models of disease or regeneration often lack relevant features to reflect the spatial and temporal events that drive fibrotic and repair processes *in vivo*. We have proposed a series of engineered tissue platforms that incorporate the progressive elements of fibrosis and related ECM heterogeneity to increase the physiological relevance of our models.

At a fundamental level, our focus is on our ability to independently and simultaneously control progressive stiffening and heterogeneity in ECM architecture and stiffness to embrace the collective effects of fibrosis. We primarily use natural biomaterials (e.g., type I collagen, hyaluronic acid) to explore techniques such as photopolymerization to control ECM stiffness over time and macromolecular crowding to control architecture. In both approaches, we are interested in bulk and local ECM changes in biophysical properties and how cells sense and respond to those changes. When considering our platforms in the context of disease, we focus on pancreatic ductal adenocarcinoma (PDAC) and lymphedema, two diseases with significant fibrotic mechanisms that influence disease progression and treatment. For PDAC, we use biomaterials that reflect *in vivo* tissue composition and focus on how progressive stiffening and heterogeneous architecture drive metastatic events and immunosuppression. We believe that the fibrotic tumor microenvironment drives those events by promoting invasive cell phenotypes and regulating cancer and stromal cell trafficking through lymphatic vessels. In lymphedema studies, we seek to improve models of lymphatic growth by decoupling system responses (e.g., chemokine signaling) from fibrotic elements (e.g., progressive stiffening) to resolve discrepancies between *in vitro* and *in vivo* results and gain insight into therapeutic approaches to restore lymphatic function. Finally, we have entered into a collaboration to explore opportunities in additive manufacturing of biodegradable metal-polymer composites that can be used to improve our understanding of processes in bone remodeling and regeneration. Moreover, we see these additively manufactured platforms as a potential way to resolve fundamental questions about the interplay between pore structure (e.g., size, shape, distribution, gradient), surface topography, and mechanics in regulating the behaviors of cell types involved in various stages of bone regeneration.

With a combination of controlled manipulation of biomaterial properties cell-mediated events, and fabrication techniques, we aim to generate a portfolio of approaches by which we can generate context-specific engineered tissues. This approach to improve *in vitro* model design will increase our overall understanding of the fundamental role of fibrosis in a host of situations. Moreover, successful models will provide further insight into how diseases develop and progress and how to control regenerative processes, particularly at the interface between repair and fibrosis, for improved patient care.

625536 Biofilm Extracellular Matrix Protein: From Functional Amyloids to Versatile Biomaterials

Abstract:

Biofilms consist of bacteria embedded in a three-dimensional (3D) extracellular matrix (ECM). An essential component of the biofilms ECM are amyloid fibrils formed by self-assembly of bacterial proteins. The functions of both protein monomers and fibers are essential for the formation, survival, and proliferation of bacterial communities. Recent insights into bacterial biofilm matrix structures have shown the importance of amyloid fibers as common building block structures that confer stability to the ECM. Amyloid fibers formed by from phenol-soluble modulins (PSMs) in *Staphylococcus aureus* (*S. aureus*) also attracted great attention in the community. These fibers display extreme amyloid polymorphism and play diverse roles in pathogenicity. We utilized the multiscale/multidisciplinary approaches merging computational methods, microscopy, spectroscopy, and microfluid, converging on 3D molecular architecture and functions of *S. aureus* amyloid proteins, PSMs, within the same range of reconfiguration times and spatial dimensions. Furthermore, we investigated their assemblies under physiological conditions including fluid flow and shear stress, and how their assembly alternate multicellular organization in the scale of complex tissues as biomimetic biomaterials. These findings led us to the design and discovery of versatile biomaterials effectively mimicking the nanoscale components of ECM and interfaces using bio-inspired building blocks. Lastly, the convergent approaches and multidiscipline efforts provide a unique platform to develop, implement, and iteratively improve integrated design protocols applicable to new materials and devices, aiming to advance their applications in healthcare.

635240 Immuno-engineered Biomaterials Reduce Implant-related Infections

Abstract:

The invention and use of antibiotics are given credit for the significant reduction of infections and saving millions of lives. However, the wide use of antibiotics has also created the latest grave challenge in patient care, i.e., antibiotic resistant bacteria, which have been increasing in recent decades. Besides antibiotic resistance, infection treatments face other challenges including biofilm formation, intra-cellular infection, delayed wound healing, etc. The recent COVID-19 have further reminded us of the challenges in dealing with infections. In this paper, engineering of biomaterials with cytokines to tune the immune responses against implant-related infections will be presented, and a few nanotechnology approaches will be briefly discussed.

621962 4D Printing: The Promise of Externally Controlling Internal Biomaterial Shape

Abstract:

4D printing has emerged as a revolution in biomaterial design and use. Specifically, 4D printing refers to materials which are 3D printed with materials than can be externally controlled to change shape. 4D printed materials can be useful for controlled drug delivery or even changing shape to increase pressure on juxtaposed tissue to promote tissue growth (such as bone). This presentation will cover some of the more exciting advances in 4D printed materials and their applications in regenerative medicine, anti-cancer applications, anti-infection strategies, and controlled drug delivery. It will also highlight what is needed for this field to grow and mature into clinically-used technology.

633800 Engineering in Precision Medicine

Abstract:

Engineered materials that integrate advances in polymer chemistry, nanotechnology, and biological sciences have the potential to create powerful medical therapies. Dr. Khademhosseini is interested in developing 'personalized' solutions that utilize micro- and nanoscale technologies to enable a range of therapies for organ failure, cardiovascular disease and cancer. In enabling this vision, he works closely with clinicians (including interventional radiologists, cardiologists and surgeons). For example, he has developed numerous techniques in controlling the behavior of patient-derived cells to engineer artificial tissues and cell-based therapies. His group also aims to engineer tissue regenerative therapeutics using water-containing polymer networks called hydrogels that can regulate cell behavior. Specifically, he has developed photo-crosslinkable hybrid hydrogels that combine natural biomolecules with nanoparticles to regulate the chemical, biological, mechanical and electrical properties of gels. These functional scaffolds induce the differentiation of stem cells to desired cell types and direct the formation of vascularized heart or bone tissues. Since tissue function is highly dependent on architecture, he has also used microfabrication methods, such as microfluidics, photolithography, bioprinting, and molding, to regulate the architecture of these materials. He has employed these strategies to generate miniaturized tissues. To create tissue complexity, he has also developed directed assembly techniques to compile small tissue modules into larger constructs. It is anticipated that such approaches will lead to the development of next-generation regenerative therapeutics and biomedical devices.

633375 Biomaterials That Enable and Drive Scaffold-Free Tissue Formation

Abstract:

Scaffold-free cultures of cells can mimic immature condensates that initiate many developmental and healing processes. Providing specific soluble signals, such as growth factors, exogenously in tissue culture media to regulate cell behavior and promote new tissue formation may be limited by transport issues, lack of spatial control over signal presentation, and required repeated dosing. The first part of this talk will be a report on systems that incorporate biomaterial microparticles containing bioactive signals within scaffold-free cell cultures, permitting localized spatiotemporal control over the presentation of these regulatory cues to the cells to engineer a variety of tissues. Three-dimensional bioprinting has been pursued as a method of building functional scaffold-free tissue constructs with sophisticated geometries without an intervening scaffold that can interfere with critical cell-cell interactions. However, to date it has not been possible to print individual cells directly, as preculturing was required to first form cell aggregates or strands prior to printing. The latter part of this talk will cover the first bioprinting platform using single cells alone, without a macromer solution, as bioink that can maintain printed geometries. A shear-thinning and self-healing biomaterial support bath was engineered to permit high-resolution printing into precise geometries, maintenance of cell viability, cellular condensation formation and long-term culture of constructs for development of engineered tissues. Collectively, these platforms may advance cellular condensation-based regenerative medicine strategies, enhance drug-screening capabilities and provide new tools for addressing questions in developmental biology.

633713 Engineering targeted therapeutics for triple negative breast cancer

Abstract:

My lab engineers targeted drug delivery vehicles in an effort to alter biodistribution and alter cell signaling. Cells organize proteins and lipids to govern adhesion and migration. From this knowledge, my lab synthesizes drug delivery vehicles to achieve strong, cooperative binding. These strategies are employed in inflammation and triple negative breast cancer. My lab has identified a new target and biomarker for triple negative breast cancer, examined the role of ligand/receptor cell adhesion by atomic force microscopy, and synthesized targeted drug delivery vehicles that demonstrate that nanoparticle ligand surface density alters gene expression.

623503 Biodegradable Scaffolds for Large Peripheral Nerve Defect Regeneration: Electrical and Chemical Stimulation

Abstract:

Every year, more than half a million Americans suffer from peripheral nerve injuries (PNI) that require surgical treatments totaling close to $2 billion in healthcare costs, but many continue to experience pain and/or poor functionality [1]. These severe PNI are typically characterized by large gaps >5 mm that are unable to regenerate on their own without surgical intervention. The main treatments for PNI involve surgical implantation of autografts or allografts, but these are frequently limited in availability and potentially cause immunosuppression, donor site morbidity, scarring, neuroma formation, and sensory loss. Tissue engineering (TE) approaches for nerve repair and regeneration, which use scaffolds, cells, and/or growth factors, can be effective alternatives to biological grafts; however, due to limits on the size of the nerve gap that can be repaired, biocompatibility, and inefficient stimulation of nerve regeneration, they are often not used or only used in specific cases. Ineffective PNI repair – a potential outcome for all current treatments can cause severe loss of sensory and/or motor function, and more severe damage is associated with slower recovery, highlighting an unmet need for better treatments. Chemical and electrical stimulation have the ability to enhance endogenous nerve regeneration in less severe PNI injuries, and may have utility in severe PNI repair [2].

Electrical stimulation (ES) as a physical therapy modality has shown benefits in the functional rehabilitation of muscle, bone, skin, tendons, ligament, and nerve injuries (2-4). Specifically, ES enhances cell proliferation, extracellular matrix synthesis, cytokine production, and vasculature development to promote overall wound healing. However, the underlying mechanisms driving these functions are poorly understood. ES of injured peripheral nerves can accelerate axonal regeneration and functional recovery in laboratory animals and human clinical trials; and accelerated axonal regeneration is correlated to higher endogenous expression of BDNF neurotrophin family growth factors and its high-affinity receptor TrkB. Furthermore, although a few clinical trials and animal studies involving ES at low to moderate treatment intensity demonstrated reductions in inflammatory and immunological responses through altered cytokine or chemokine release and/or cell activation, survival, proliferation, or death, ES effects on implant immune responses have not been studied. Additionally, the full potential of ES-mediated tissue regeneration in large-area tissue defects including large-gap PNI is not known, and challenges exist in the application of reliable stimulation to achieve consistent healing outcomes for nerve crush injuries. One challenge with applying ES in vivo is that ES signal transduction from dermis nerves to skin-surface electrodes is not reliable for stimulation or measurement in the absence of suitable connecting materials. Nonetheless, a recent expert article describes the use of ion-conducting hydrogels as an electrode interface that improves signal transduction to deliver reliable ES to nerves, which may be less painful for long-term application. Therefore, more comprehensive and in-depth studies are needed to enhance regenerative capacity and functional outcomes in large-gap PNI.

Chemical stimulation with the potassium-channel blocker 4-aminopyridine (4-AP) can restore the myelination of demyelinated nerves in multiple sclerosis – most likely through its ability to prolong action potentials and amplify neurotransmitter release; myelin ensheathes axons and acts as an electrical insulator, greatly speeding up action potential conduction [3]. Application of 4-AP to crush nerve injuries – where axons are intact but the myelin sheath is damaged – can also increase remyelination and nerve conduction velocity, leading to functional recovery. Therefore, 4-AP has potential to enhance large-gap PNI repair, but local application in large-gap PNI and its effects on regeneration and functional recovery have not been studied.

Delivering growth factors or cells with scaffolds can stimulate nerve regeneration processes, but chemical and electrical stimulation may have similar and additional benefits with fewer limitations [1]. Neurotrophins such as nerve growth factor (NGF) and brain-derived neurotrophic factor (BDNF) can improve nerve regeneration, but proteins can have high dose requirements, short half-lives, high costs, and undesired side effects. Schwann cells and other neuronal support cells can accelerate nerve regeneration as well, but delivering viable cells can be cumbersome and costly. No efforts have been made to encourage infiltration of endogenous Schwann cells using a drug or any other external cue to improve large-gap PNI. ES and 4-AP may accomplish this goal but will require innovative design of new polymers and delivery methods.

Ionically conductive (IC) polymers and sulfonated chitosan especially, confer high and sustained conductivity, overcoming limitations of other materials in ES applications [4,5]. IC polymers conduct electrical charges via the flow of counter ions in the physiological environment, which is different from electronically conducting polymers, and can be synthesized to carry positive (cation) or negative (anion) charges from well-studied biodegradable polymers, such as chitosan, silk, and poly(lactic acid). Degradation mechanisms and rates vary by polymer type, and include enhanced surface erosion via ionic group modification-conferred hydrophilicity of the polymer. We have developed sulfonated IC polymers with excellent tunable electrical conductivity for neural tissue engineering, and found that sulfonated chitosan has higher initial and sustained conductivity than other sulfonated and non-sulfonated polymers. The non-degradable sulfonated polymers poly(ether ether ketone) (S-PEEK) and poly(aniline) (S-PANI) also showed sustained conductivity.

Results and Discussion: Our scaffold system is based on cross-linked sulfonated (S) chitosan, a natural sugar polymer found in crustaceans and fungi, and designed with: high-sustained conductivity critical for long-term ES; strength and flexibility for kink resistance and suturability; aligned pores for cell guidance; and nanotubes for controllable small molecule release. We developed a freeze-drying technique to create cross-linked S-chitosan IC scaffolds with longitudinally aligned pores optimal for maximum axon penetration and minimum axon misdirection [3]. These also accommodated and guided Schwann cells needed for myelination of axons (data not shown). In our preliminary scaffold formulation, 600–900 nm halloysite (alumosilicate clay) nanotubes (HNT) were filled with 4-AP and uniformly dispersed in IC chitosan solution to fabricate porous scaffolds. These IC scaffolds are kink-resistant and suturable, with suture retention strength ranging 18–20 N. Drug-filled HNTs are sealed with chitosan at both ends before combining with chitosan solution. By varying HNT content and cross-link density, we can tune the architecture to match native peripheral nerve mechanical modulus and control drug release. The ability to control localized physical and chemical signals with scaffolds will allow extended treatment periods to achieve motor function not typically seen with short-term treatments.

4-AP induces dose-dependent increases of neurotrophins in human Schwann cells in vitro. Human Schwann cells cultured for 14 days with varying doses of 4-AP induced expression of NGF, P0, and BDNF neurotrophins. These neurotrophins promote axon regeneration, remyelination, and motor function recovery. IC scaffolds support human Schwann cell adhesion, proliferation, and neurotrophin secretion in response to 4-AP and ES in vitro. We published that IC scaffolds can support Schwann cell adhesion, proliferation, and migration along aligned pores, as well as secretion of neurotrophins in response to 4-AP. Intriguingly, our new preliminary data suggest that compared with ES or 4-AP alone, combined 4-AP+ES treatment increases expression of neurotrophins BDNF, NGF, and S100 in

Schwann cells grown on IC scaffolds in vitro at day 14 [3].

Combined ES+4-AP treatment improves foreign body responses compared to IC scaffolds alone in vivo. To test whether ES, 4-AP, and/or 4-AP+ES elicit an immune response compared to scaffolds alone, we implanted scaffolds subcutaneously in rats, injected IC scaffold formulations as injectable electrodes to deliver ES treatment groups, and applied ES every other day. Blood samples from tail veins were collected several times over 14 days to analyze IL-6 and IL-10 cytokines by ELISA. Levels of IL-6 and IL-10 fluctuated over time (not shown), and the IL-6/IL-10 ratio – a measure of immune response – dropped initially and stabilized thereafter indicating no strong immune response to implanted scaffolds over 14 days. Fiber capsules can form around implanted foreign material – another measure of immune response – therefore, we measured scaffold fiber capsule thickness using H&E-stained sections. Reduced fiber capsule thickness is an indicator of reduced foreign body response and enhanced host-tissue integration.

IC scaffolds+4-AP and autografts confer similar early-stage sciatic nerve defect repair. Our preliminary data using 4-AP-loaded chitosan-HNT (scaffolds+4-AP) in this same 15 mm nerve defect model showed repair responses (nerve histology, myelin production) and functional recovery (sciatic functional index; SFI) comparable to autografts over 8 weeks[6]. Notably, scaffolds+4-AP appeared better histologically, but similar functionally, than scaffolds alone at this early stage of nerve repair. Thus, investigating long-term 4-AP effects, safety, and full functionality (beyond 8 weeks) is critical. Autografts had the thickest myelin (G-ratio) followed by scaffolds+4-AP, and scaffolds alone. Myelin content in autografts, scaffolds, and scaffolds+4-AP were lower than sham (surgery) controls, however, there were no differences between test groups. Neurofilament-H (NF-H) and S100 staining was observed for all groups, with sham control exhibiting highest and scaffolds alone exhibiting lowest content. In addition, 4-AP application appeared safe and bierosion products didn't indicate any toxicity in liver histology[6].

Conclusions: Delivering growth factors or cells with scaffolds can stimulate nerve regeneration processes, but chemical and electrical stimulation may have similar and additional benefits with fewer limitations. This facilitates more efficient and efficacious peripheral nerve regeneration via a drug delivery system that is feasible for clinical application.

633802 Advances in translation of SAPH peptide biomaterial-drugs

Abstract:

Self-assembling peptide hydrogels (SAPH) are a class of injectable scaffolds that present a paradigm in drug development and biomaterials owing to their tunable specificity and simple manufacturing. Facile self-assembly of monomeric/ multi-meric constituents result in high epitope presentation of biological signals. Persistent signaling, in situ bolus delivery and demonstrable modification to actuate specific biological responses allow development of novel classes of biomaterials that behave as scaffolds and drugs. This presentation will describe a few examples of novel peptide-biomaterial drugs that are capable of site-specific delivery to tissue, potentiating numerous tissue remodeling responses. A focus is also made to describe entrepreneurship and innovation helping bridge the gap between academia, industry and clinical medicine. We begin by describing the platform and some interesting applications to-date, efforts made todays translation and commercialization, and challenges therein. We further describe some recent work targeting and inhibiting SARS-CoV-2 for applications in our current pandemic.

633782 Recent Advances with Silk Proteins for Medicine

Abstract:

Silk is one of the oldest biomaterials, utilized as sutures for centuries, yet undergoing a rebirth into new biomaterial formats and applications. Fundamental aspects of control of silk protein biomaterial structure and function have led to new silk-based biomaterials for drug delivery, tissue engineering and regenerative medicine. Exploitation of silk processing, along with specialized functionalization, can be used to modulate biological responses, mechanics, degradation lifetime and biological utility in vitro and in vivo. Recent advances in technological innovation and translation of silk biomaterials will be discussed.

AUTHOR INDEX

Alsberg, Eben 8
Auguste, Debra 9
Jiang, Shaoyi 1
Kaplan, David L 14
Khademhosseini, Ali 7
Kumar, Vivek 13
Kumbar, Sangamesh 10
Li, Bingyun 5
Wang, Yichun 4
Webster, Thomas 6
Whittington, Catherine 3
Zhang, Y. Shrike 2

AIChE
120 Wall Street, FL 23
New York, NY 10005-4020

ISBN 978-1-7138-5688-7

9 781713 856887

CURRAN ASSOCIATES INC.
proceedings.com